Women on t
Vol. .

Journeyscapes of Domestic Violence

Janet C. Bowstead

Dakie Publishing

https://www.womensjourneyscapes.net

Women on the Move – Vol 1: Journeyscapes of Domestic Violence

Copyright © 2021 Dakie Publishing

All rights reserved. No part of this publication may be reproduced or transmitted in any form or by any electronic or mechanical means, including information storage and retrieval systems, without prior written permission from the author, except in the case of a reviewer, who may quote brief passages embodied in critical articles or in a review.

Published by Dakie Publishing, London
Design and Content edited by Janet C. Bowstead
Images by Amy, Carol, Cordelia, Daisy, Favour, Kate, Kelly, Lee, Lucy, Lulu, Marilyn, Marita, Qiana, Sarah, Shalom and Violet

ISBN: 978-1-7399686-0-1

Printed and bound by Cambrian Printers Ltd, Tram Road, Pontllanfraith, Blackwood, NP12 2YA

Key Messages

- Tens of thousands of women and children within the UK are forced to relocate due to domestic violence every year, as an internal displacement caused by the human rights violation of violence against women
- Beyond the force of the abuser, domestic violence journeys are further forced by policy, practice and law creating a hostile terrain that limits women's options and agency
- Women face an unknown landscape for their journeys and need and deserve route maps, signposts, and roadside assistance – services allowing, enabling and assisting women's own strategies – so that they can go as far as they need and stay as near as they can
- Women's help-seeking journeys may be functionally segmented – to escape the abuse and reach a suitable place – but are further fragmented by policy and practice; whether they stay put, remain local or go elsewhere
- The state should be journeyscaping routes – creating a coherent and accessible infrastructure for women and children's rights to safety and freedom
- The journeys may be hidden on the individual scale – due to the risk of abuse – and hidden at the local authority scale – because of the lack of net effect – but they do not need to be so unknown at the aggregate scale to decision-makers; as they often leave administrative traces, including in the extensive administrative data of service monitoring
- Administrative data from service monitoring records can be used to provide much larger samples than from survey or qualitative methods – and crucially so for women and children on the move who are excluded from the sampling frame of extensively used surveys such as the Crime Survey
- Currently, these data are often kept (and/or deleted) within organisational and geographical boundaries due to the competitive contractual environment of service provision
- In contrast, data needs to cross administrative boundaries to reveal geographical patterns and to be de-identified and archived to provide evidence over time
- Boundary-crossing data reveal that the scale of women's help-seeking is further than within local authorities; and therefore that planning and providing services at the local authority scale will not uphold women's and children's rights nor meet their needs
- Women on the move due to domestic abuse provide insights into the infrastructure needed to tackle the causes and consequences of their displacement

Dedication

To all the women who shared their experiences, insights and images for this research; and to the wider community of women and children on the move due to domestic violence and abuse. In the hope that you can see and know the strength and achievements of those who have journeyed before you, and believe that you too deserve better.

Acknowledgements

This is a publication from the research project "Women on the Move: the Journeyscapes of Domestic Violence". The research was funded 2016-2021 by the British Academy (grant number PF160072) and based at the Geography Department, Royal Holloway, University of London. The participatory photography groupwork was carried out with the specialist domestic violence service provider Solace Women's Aid. Thanks to all these contributions, and especially to the women who shared the copyright of their images for use in this publication. The author remains solely responsible for the content of this publication, and apologises for any errors and omissions.

Images copyright

© Amy/Solace Women's Aid/Janet Bowstead
- pages 17, 37, 38, 40, 41, 44, 53, 56, 71, 72, 79, 84, 85, 90, 97

© Carol/Solace Women's Aid/Janet Bowstead
- pages 13, 20, 27, 29, 30, 32, 36, 38, 41, 57, 69, 89, 92, 96

© Cordelia/Solace Women's Aid/Janet Bowstead
- pages 12, 19, 22, 33, 34, 35, 36, 40, 41, 48, 52, 54, 60, 63, 78, 79, 90, 95, 99

© Daisy/Solace Women's Aid/Janet Bowstead
- Chapter heading and pages 13, 19, 22, 27, 30, 31, 36, 38, 39, 45, 47, 48, 51, 52, 54, 57, 59, 71, 75, 76, 79, 81, 82, 84, 91, 92

© Favour/Solace Women's Aid/Janet Bowstead
- pages 27, 46, 89

© Kate/Solace Women's Aid/Janet Bowstead
- pages 26, 54, 69, 74, 89, 90, 98

© Kelly/Solace Women's Aid/Janet Bowstead
- pages 12, 18, 30, 38, 60, 68, 82, 88

© Lee/Solace Women's Aid/Janet Bowstead
- pages 15, 16, 89

© Lucy/Solace Women's Aid/Janet Bowstead
- pages 14, 16, 25, 26, 30, 34, 35, 43, 60, 68, 76, 92, 93, 94

© Lulu/Solace Women's Aid/Janet Bowstead
- pages 15, 54, 74, 89

© Marilyn/Solace Women's Aid/Janet Bowstead
- pages 18, 19, 22, 27, 38, 71, 75, 76, 79, 82, 83, 84, 91

© Marita/Solace Women's Aid/Janet Bowstead
- pages 13, 21, 23, 30, 59, 62, 71

© Qiana/Solace Women's Aid/Janet Bowstead
- pages 20, 23

© Sarah/Solace Women's Aid/Janet Bowstead
- Back cover and pages 11, 13, 14, 15, 16, 18, 20, 34, 44, 50, 62, 63, 76, 78, 81, 88, 90

© Shalom/Solace Women's Aid/Janet Bowstead
- Front cover and pages 13, 16, 18, 23, 27, 30, 36, 45, 55, 57, 61, 66, 71, 74, 82, 87, 90

© Violet/Solace Women's Aid/Janet Bowstead
- pages 30, 33, 88, 92, 100

Contents

- Key Messages 3
- Dedication & Acknowledgements 4
- Images copyright 5
- Introduction 7
- **PEOPLE** 9
  - Rights — 10
  - Women – Demographics — 14
  - Women – Meanings — 17
  - Strategies – Options and Agency — 21
  - What about the men? — 24
  - What about the children? — 25
- **PLACES** 28
  - Different types of places – Rural/Urban — 29
  - Rates of help-seeking from different places — 31
  - Getting there – Transport — 34
  - Regions — 37
  - London — 39
- **PATTERNS** 42
  - Distance — 43
  - Journey trajectories – the start — 46
  - Journey trajectories – stages — 49
  - Journey trajectories – time — 52
  - Journey trajectories – multiple displacements — 55
  - Force and Agency — 58
  - Pressure Points — 61
- **PROCESSES** 64
  - State Duties – Admin data as evidence — 65
  - State Duties – Journeyscapes — 67
  - State Duties – Laws — 70
  - State Duties – IDPs — 73
  - Strategies – Scale — 77
  - Strategies – Boundaries — 80
  - Strategies – Referrals and Support Services — 83
  - Strategies – Women's Refuges — 86
  - Strategies – Housing and Home — 91
  - Formula for service provision — 93

Introduction

This research investigates women's domestic violence journeys – their journeys to escape an abusive partner – at a range of scales from individual to local, national and international. Such journeys are necessarily hidden; but the research has generated a wealth of evidence from de-identified administrative data from services in England, and insights and knowledge via creative participatory groupwork with women in London who have relocated due to abuse. It provides a new conceptualisation of women's domestic violence journeys as a forced internal migration caused by the human rights violation of violence against women.

This publication provides an overview of the project, highlighting key research findings and new conceptualisations and knowledge from the research. These have also been presented and published in academic journals, practitioner reports and briefing papers, at conferences and policy events online and offline during the project (2016-2021).

More than anything, it provides an opportunity to present some of the images from the groupwork (selected out of over 1,500 images) by women who have been forced to relocate due to domestic violence, highlighting their creativity and insights as women on the move. Nineteen women took part in three groups in different areas of London and sixteen put photographs into the research: Amy, Carol, Cordelia, Daisy, Favour, Kate, Kelly, Lee, Lucy, Lulu, Marilyn, Marita, Qiana, Sarah, Shalom and Violet (their chosen pseudonyms).

The research has therefore achieved greater recognition and understanding of the extent of forced relocation of women and children in the UK, and the experiences of tens of thousands of women on the move. The research has largely achieved its aim of developing a multi-scaled understanding of both the processes of women's domestic violence journeys, and their implications, and this publication is a further stage of presenting this to a range of policy, practice, academic and public audiences.

Much still needs to be done

- to shift policy and political thinking in terms of understanding this gendered process of forced displacement and acceptance of these women and children as Internally Displaced Persons (IDPs), with the consequent duties of the UK state to minimise their displacement, support their retention of rights and possessions, and assist their resettlement.
- to champion the research use of administrative data from services: such monitoring data as that used extensively in this research are still usually inaccessible for research due to commercial concerns in the competitive tendering environment and the lack of de-identification and archiving of data.
- to act on the research findings on the necessary infrastructure for women and children's needs for rights and services for both relocation and staying put. The specific output of a formula for rights- and needs-based service provision – location and capacity – has had little purchase so far in the current political and service provision context.

The following sections highlight key messages – and point to further reading – on aspects of People, Places, Patterns and Processes of women and children on the move due to domestic violence and abuse.

Further details are available via the project website:

https://www.womensjourneyscapes.net/

Rights

It shouldn't be too controversial a statement to highlight that violence against women is not just an individual problem – causing fear, harm and injury – but a human rights violation. A violation that does not just harm the individual woman, but harms society, community, nation and humanity – every time that abuse is not responded to with justice.

Violence against women does not just need a response from welfare services to individuals; it needs a response from justice services in the widest sense. It needs a willingness to tackle the causes as well as the consequences of the abuse.

That is the thinking about linking the International Day for the Elimination of Violence against Women (25th November) with International Human Rights Day on the 10th December via the internationally-recognised "16 days of activism" each year[1].

So, violence against women is recognised internationally as a human rights violation:

> *"the elimination of violence against women in public and private life is a human rights obligation"* – *"this form of violence impedes the ability of women and girls to claim, realize and enjoy their human rights on an equal foot with men."*[2]

And human rights violations are one of the recognised causes of displacement – whether internationally (leading to refugees) or within countries (leading to Internally Displaced Persons – IDPs).

The UN Office for the Coordination of Humanitarian Affairs (UNOCHA) publishes the "Guiding Principles on Internal Displacement", highlighting the distinctive issues around internal displacement:

> *"Unlike refugees, the internally displaced have not left the country whose citizens they normally are. As such, they remain entitled to the same rights that all other persons in their country enjoy. They do, however, have special needs by virtue of their displacement."*[3]

References
[1] https://www.unwomen.org/en/what-we-do/ending-violence-against-women/take-action/16-days-of-activism
[2] Office of the High Commissioner for Human Rights (UN Human Rights) https://www.ohchr.org/EN/Issues/Women/Pages/VaW.aspx
[3] *Handbook for applying the Guiding Principles on Internal Displacement* (UN Office for the Coordination of Humanitarian Affairs, 1999, p. 5) http://www.unhcr.org/en-us/protection/idps/50f94df59/handbook-applying-guiding-principles-internal-displacement-ocha-november.html

The UNOCHA specifically lists human rights violations as one of the causes of internal displacement:

> "The reasons for flight may vary and include armed conflict, situations of generalized violence, violations of human rights, and natural or human-made disasters." [1]

Therefore, it is clear that internal displacement due to violence against women is within these definitions. That forced displacement of women and children due to domestic violence creates internally displaced persons (IDPs). However, this is not currently properly recognised in terms of women's and children's rights – and state responsibilities to uphold those rights and meet their needs.

Women themselves know that we/they deserve better.

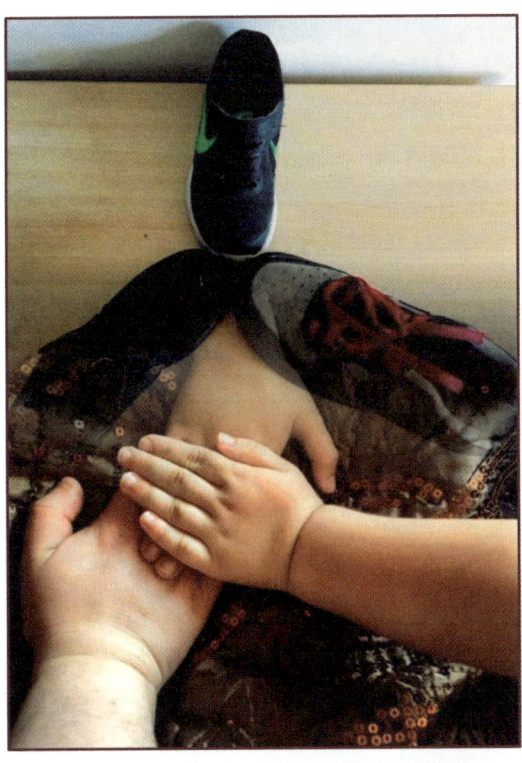
© Sarah

"When four becomes three! You can't walk over us no more. We're in power because 3 Beats 1"

Sarah

Reference
[1] *Handbook for applying the Guiding Principles on Internal Displacement* (UN Office for the Coordination of Humanitarian Affairs, 1999, p. 5) http://www.unhcr.org/en-us/protection/idps/50f94df59/handbook-applying-guiding-principles-internal-displacement-ocha-november.html

© Cordelia

"There will be rainy teary days, but the sun will always come out again"

Sarah

"I love – when they swim – what they do to the water – all the patterns that happen"

© Kelly

"Just wanted to share how it makes me feel and it's freedom"

Kelly

© Daisy

© Marita

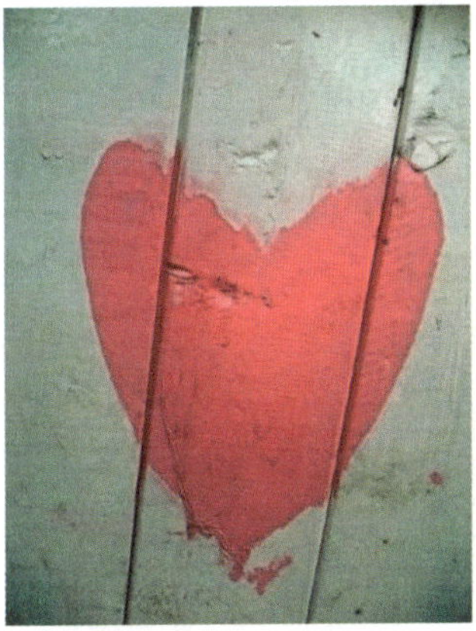

© Sarah

© Sarah

"Because even in the most unexpected places there is life, love and light..."

Sarah

© Shalom   © Carol   © Shalom

Women - demographics

Each woman's journey to escape domestic abuse is unique, and only some include formal services in their help-seeking. However, women of all ages and backgrounds, with and without children, seek help. Women aged 15 – 102, from all ethnic backgrounds and religions, are recorded as accessing services.

© Lucy

© Sarah

© Lucy

"Tea, I start every day with a tea" — **Lucy**

Further Reading

Bowstead, Janet C. 2015. "Forced Migration in the United Kingdom: Women's Journeys to Escape Domestic Violence." *Transactions of the Institute of British Geographers* 40 (3): 307–320. doi:10.1111/tran.12085.

Bowstead, Janet C. 2021. "Stay Put; Remain Local; Go Elsewhere: Three Strategies of Women's Domestic Violence Help Seeking." *Dignity: A Journal of Analysis of Exploitation and Violence* 6 (3): 4. doi:10.23860/dignity.2021.06.03.04.

https://www.womensjourneyscapes.net/wp-content/uploads/2021/07/Womens-Journeyscapes-Briefing-paper-7-July-2021.pdf

https://www.womensjourneyscapes.net/wp-content/uploads/2021/10/Womens-Journeyscapes-Briefing-paper-8-Oct-2021.pdf

© Lulu

"My Bible, the White Roses, My Lord, Virgin Mary & Holy Spirit"

Lulu

"My religion"

Lee

© Lee

© Sarah

"Home-made with love and passion for my children. Albanian dish "Byrek" – spinach and feta cheese pastry"

Sarah

Women - meanings

Women's narratives of their domestic violence journeys highlight practical, emotional and legal issues for women on the move; but also the impact on their sense of displacement and identity for themselves and their children. Women's voices are presented in quotations on the research website and here their images and captions enable understandings of their senses and meanings of home, identity and belonging in their experiences of displacement and resettlement.

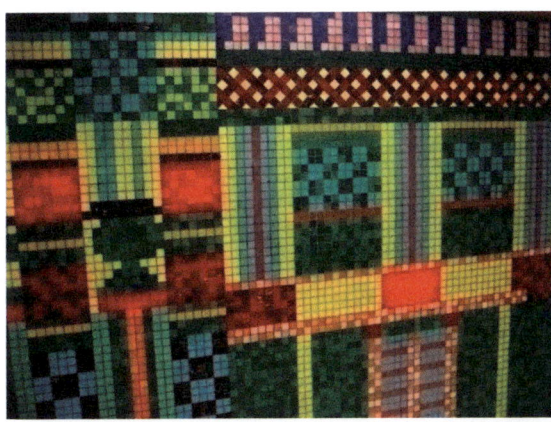

© Amy

"Those repeating tile patterns have the ability right now to disorientate me – it broke my balance system with the stress – So when it was really bad, I would be literally stopped in my tracks" **Amy**

Further Reading

https://www.womensjourneyscapes.net/becoming-yourself-again/

https://www.womensjourneyscapes.net/moving-and-moving-on/

https://www.womensjourneyscapes.net/imagined-community-real-belonging/

https://www.womensjourneyscapes.net/going-back/

Bowstead, Janet C. 2021. "There Is Always a Way Out! Images of Place and Identity for Women Escaping Domestic Violence." In *Representing Place and Territorial Identities in Europe: Discourses, Images, and Practices*, edited by Tiziana Banini and Oana-Ramona Ilovan, 191–202. Cham, Switzerland: Springer. https://link.springer.com/chapter/10.1007/978-3-030-66766-5_13.

© Kelly

"A path leading somewhere – somewhere – it's still quite dark, but we'll see where it goes!"   **Kelly**

"The tree – struck by lightning – but it carries on"   **Shalom**

© Shalom

© Sarah

"Sometimes one is better than a crowd – Being free – feeling good – not lonely"   **Sarah**

"I love this tree – it's been chopped down – but all the shoots have started to grow again – like a magical tree"   **Marilyn**

© Marilyn

© Daisy

"It's like an abandoned chariot – to make a story somehow – the wheels have come off"

Daisy

"I'm always on the move – and sometimes I think I'm running away from myself – I think – why can't you sit still? For 5 minutes"

Cordelia

© Cordelia

© Marilyn

"I just thought it's like weird. . You know – like mixed wires – mixed signals – getting your wires crossed!"

Marilyn

© Sarah

"From black and white to colour! How life changes very quickly!"

Sarah

©Carol   © Qiana
© Qiana   © Qiana

Strategies – options and agency

Each woman's journey to escape domestic abuse is unique, but there are shared characteristics that can be identified. Understanding these shared aspects can help identify the barriers and challenges that face particular women or women in particular places or circumstances. This can highlight the implications of domestic violence journeys and the kinds of support needed from different types of services, laws or policies.

The data from women who did access services show three distinct strategies when women seek help:

➢ Stay Put – and seek help from support services
➢ Remain Local – relocate to access support, but within the same Local Authority
➢ Go Elsewhere – move to another Local Authority to seek help

© Marita   © Marita

"I was treated like rubbish!" Marita

Further Reading
Bowstead, Janet C. 2021. "Stay Put; Remain Local; Go Elsewhere: Three Strategies of Women's Domestic Violence Help Seeking." *Dignity: A Journal of Analysis of Exploitation and Violence* 6 (3): 4. doi:10.23860/dignity.2021.06.03.04.

© Daisy

© Marilyn

© Daisy

"Thinking 'which way should I go?' – that's so much of my life – which way should I go?"

Cordelia

© Cordelia

© Shalom

"Aloe Vera – I use that for everything – it's really good for hair, especially for your scalp – it's so soothing: I put it on my hair and on my scalp. And I use it to make Aloe Vera oil – for my skin and for cuts and bruises – it's really healing"

Shalom

© Marita

© Qiana

© Marita

© Marita

"Make some noise!"

Marita

What about the men?

Further Reading
https://www.womensjourneyscapes.net/wp-content/uploads/2018/08/Womens-Journeyscapes-Briefing-paper-2-June-2018.pdf

What about the children?

Over half the women making journeys to services have children with them – and their children are important in what they are thinking and doing. Mothering on the move is not easy – women are trying to support their children to make sense of the abuse and to understand the need to relocate; at the same time as dealing with all the practical and emotional disruptions.

Women often have very little choice about when and where they seek help – both because of the threat of the abuser, and the lack of service options. This includes the fact that many mothers of school-age children cannot avoid relocating during term-time, and children often face a further wait to get into a new school – and still longer to settle and begin to catch up. However, in escaping the violence, mothers also talk about the positives as their children are able to feel safe and begin to recover; and they themselves are able to be the mother they would want to be.

"Safe from the sharks circling"

Lucy

© Lucy

Further Reading
https://www.womensjourneyscapes.net/mothering-on-the-move/
https://www.womensjourneyscapes.net/children-on-the-move-what-about-their-schooling/

© Kate

"Like when we were living in the situation – before – we wouldn't necessarily go out at night-time; but now it's like freedom to do whatever you want and go out whatever time you want"

Kate

© Lucy

© Lucy

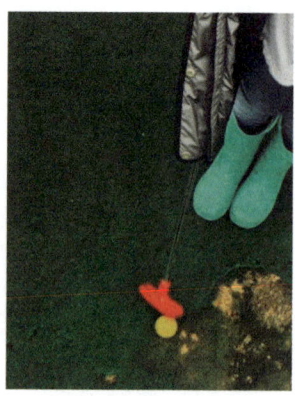

© Lucy

"Days out – golf – cinema * making memories * trying something new * coming out of your comfort zone"

Lucy

Different types of places – Rural/Urban

Women from every local authority seek service help – see the flow diagram enclosed – and access services in every local authority that provides any kind of support service. But the pattern is not of completely random flows: women were significantly more likely to go to the same type of local authority as the one they left (in terms of classifications of type of area – such as how rural or urban), and least likely to go to the most dissimilar types of places.

Public transport routes are generally cheaper and easier on the main routes into major urban areas, and more difficult, costly and infrequent in more rural areas.

So, a strong flow of women to major urban areas might be expected as women travel away from rural areas where there are limited support services and less public transport, and away from small towns where women might fear being more easily noticed as a newcomer, and easier to track down. But in fact there are not strong flows along public transport routes, and major cities are consistently places of net *leaving* due to domestic violence; with more women and children leaving each year to services elsewhere in the country, than the number who arrive to access services in the city.

Domestic violence journeys are not about really wanting to change place. It makes sense – women are trying to escape the violence, but they want to stay in the kind of place where they and their children can start again after abuse.

© Carol

Further Reading
Bowstead, Janet C. 2015. "Forced Migration in the United Kingdom: Women's Journeys to Escape Domestic Violence." *Transactions of the Institute of British Geographers* 40 (3): 307–320. doi:10.1111/tran.12085.

Bowstead, Janet C. 2020. "Private Violence/Private Transport: The Role of Means of Transport in Women's Mobility to Escape from Domestic Violence in England and Wales." *Mobilities* 15 (4): 559–574. doi:10.1080/17450101.2020.1750289.

http://www.womensjourneyscapes.net/my-kind-of-town/

Rates of help-seeking from different places

Different places have different numbers of women seeking help; whether staying put, remaining local, or going elsewhere to access service support. The administrative record of help-seeking reflects the outcomes of tens of thousands of decisions that affect where women and children go. But all those individual actions can be aggregated to consider the numbers per year and the numbers from individual local authorities.

The overwhelming factor affecting numbers is the population of the local authority – so that the rate per female population of different help-seeking strategies is also crucial to understand and provide for both women and children who stay put or remain local, and for those who go elsewhere across local authority boundaries. While rates of going do vary between places (and are considered in detail in the proposed service provision formula), the net effect – leaving minus arriving – for most local authorities is around zero.

© Daisy

> "It's been through a lot – That's some stories!"
>
> **Daisy**

Further Reading
Bowstead, Janet C. 2015. "Forced Migration in the United Kingdom: Women's Journeys to Escape Domestic Violence." *Transactions of the Institute of British Geographers* 40 (3): 307–320. doi:10.1111/tran.12085.

© Cordelia

"I was thinking that kind of sums me up at the minute because the sun is that I'm really optimistic and my life's quite good at the minute because I made a lot of changes. But then the fog in this is like I'm still a little bit unsure"

Cordelia

© Violet

Getting there - Transport

The lack of strong domestic violence journey flows along public transport routes, highlights the need to understand more about how women get to support services – what means of transport do they use?

Whilst public transport can be vital, especially for longer distances and to unknown places – for example to access a women's refuge – women use all types of transport, and often different types at different stages of their journeys.

At each stage, the type of transport can depend on the options available, the threats faced, and how much control women have over where and when they are travelling; as well as whether they can safely draw on the help of others. Disabled women may have fewer transport options, as may women from rural areas, and be more reliant on the help of others and/or their own private transport. The type of transport does not just affect the cost and time of the journey, but also whether women can retain their and their children's possessions.

© Lucy   © Cordelia   © Sarah

Further Reading
Bowstead, Janet C. 2020. "Private Violence/Private Transport: The Role of Means of Transport in Women's Mobility to Escape from Domestic Violence in England and Wales." *Mobilities* 15 (4): 559–574. doi:10.1080/17450101.2020.1750289.

https://www.womensjourneyscapes.net/transport-yourself-to-a-better-place/

© Lucy

"I spend a lot of time travelling in my car, so this is a everyday view for me"

Lucy

"That's where I used to get in coming from outside London – I just call it my Escape – I always used to end up there"

© Cordelia

© Cordelia

"When I first came to London, I didn't know anyone or how to get anywhere. So I just used to jump on any bus and see where it took me and that's how I got to know my way around"

Cordelia

Regions

Women and children's journeys across local authority boundaries – see the flow map enclosed – show that local authorities are very far from being self-contained in terms of domestic violence help-seeking. There is a pattern of spatial churn criss-crossing the country.

However, regions are much more self-contained than Tier 1 (county and unitary) authorities, varying from 93 per cent of help-seeking journeys from the North East remaining within that region, to 75 per cent from London remaining within London (as the least self-contained region). A regional approach to service responses would therefore fit better with the scale of women's help-seeking than the current localised approach.

© Amy

Further Reading
Bowstead, Janet C. (forthcoming). "Journeyscapes: the regional scale of women's domestic violence journeys."

London

London is net-leaving in terms of domestic violence help-seeking – so the region needs the rest of the country more than it serves it with support services for women and children. In other types of internal migration, London pulls people in from the rest of the country; but in this gendered and forced internal migration due to domestic abuse, more London women leave than others arrive. Some may be able to return to London after service support, and there is a massive churn of domestic violence journeys across London between different boroughs.

London is distinctive in terms of the fact that it is the only region where a pan-regional approach to some service provision has been developed in recent years; and this is due to continue in the duty to provide safe accommodation under Part 4 of the Domestic Abuse Act 2021. It is ironic that the Government has decided that the regional scale is only relevant for London, the least self-contained region in England.

© Daisy

"London calling me – This was in Camden"

Daisy

Further Reading

Bowstead, Janet C. (forthcoming). "Journeyscapes: the regional scale of women's domestic violence journeys."

https://www.womensjourneyscapes.net/wp-content/uploads/2018/08/Womens-Journeyscapes-Briefing-paper-3-July-2018.pdf

https://www.womensjourneyscapes.net/londons-churning-londons-churning/

© Cordelia

"London – by my bedside so I know where to go"

Cordelia

© Amy

© Amy

"Walking past it – what is that? – looking up and it was a weird chunk of shiny metal – looks different from whatever angle"

"Standing on Dartmouth Park hill – looking down into the city"

Amy

© Cordelia

© Amy

40

Distance

Women may travel hundreds of miles to access services; and that might be just one stage of a multi-stage journey trajectory. However, distance can be thought of both literally – geographically – and metaphorically. Journeys can be graphed over time and distance, highlighting that longer mileage does not necessarily equate with greater displacement and disruption. A long distance, but to a similar type of place (a small town, an industrial city) may end up feeling less unsettling – less displacing – than a shorter geographical distance to a completely different type of place.

Women may add up the mileage travelled, but they may also emphasise the number of moves, or the extent of control and choice – or lack of control and choice – that they have over the timing, places and distances. They may talk of how far they have travelled – but also of the dislocation even when the literal distance is short. Over half of women access services within their own local authority – staying put or travelling relatively short distances – and journeys elsewhere range from a few miles to hundreds. The principle should be that women and children go as far as they need – *but not be forced any further* – and stay as near as they can – *but not live in fear*.

© Lucy

Further Reading
Bowstead, Janet C. 2017. "Segmented Journeys, Fragmented Lives: Women's Forced Migration to Escape Domestic Violence." *Journal of Gender-Based Violence* 1 (1): 43–58. doi:10.1332/239868017X14912933953340.

© Amy

"Careful steps – My shoes and walking in the street – the old and the new"

Amy

"Taking little steps. I was thinking about permanence and places and leaving a mark that you'd been places"

Amy

© Amy

© Sarah

"Steps forward to a better life, full of love, wealth and smiles all round"

Sarah

Journey trajectories — the start

Women's journeys do not start when services are involved - so the start of their journeys from abuse to freedom will not be found in the administrative record. Women constantly develop and pursue strategies during the relationship and ongoing to manage their own lives – they were doing this before services were involved, and will be doing this long after services and professionals are involved in their lives.

Many will never access formal support services: some never hear of services or never try to contact them, and others face a dangerous period of searching and waiting for help – even being turned away. Of those women who do manage to find a place in a support service, many are not moving directly from the abusive relationship, but are already on the move. Services often only see a snapshot – a moment – but can be key in allowing, enabling or assisting women's journeys (or, conversely, blocking, blanking or breaking women's strategies).

© Favour

> "We made a good deal of use of our knapsacks in the periods we were homeless"
>
> **Favour**

Further Reading
Bowstead, Janet C. 2017. "Segmented Journeys, Fragmented Lives: Women's Forced Migration to Escape Domestic Violence." *Journal of Gender-Based Violence* 1 (1): 43–58. doi:10.1332/239868017X14912933953340.

https://www.womensjourneyscapes.net/the-opportunity-to-allow-enable-or-assist/

© Daisy

"So this is hopeful – I don't know if I'll see anyone – kind of bump into anyone. I'm just being careful carefully. I kind of tread carefully – kind of fearful"

Daisy

© Daisy

© Daisy

© Cordelia

"I think it's a message because I say – for the past week – there's one outside my flat every day. So I thought maybe it's a sign that I need to get on my bike"

Cordelia

© Daisy

Journey trajectories — stages

Whilst women may be able to identify different stages of their journey looking back, they often have little opportunity to plan – or even predict – each move at the time. They are constantly assessing and judging what might meet their needs and their children's needs within a context of lack of information and lack of options. They may be buffeted by conflicting demands, and steamrollered by agencies and professionals trying to fit them into a model of service provision or stages of recovery. They may have to make life-changing decisions without the time or the information to be able to consider the implications or any alternatives.

At every twist and turn of a woman's journey away from abuse she is balancing the threats and risks against her with her own needs and plans and options. It is a constantly changing assessment, as she tries to assess what she needs and wants, against what she can find to help her.

Meanwhile, as she interacts with services and professionals, they may be carrying out formal assessments of her — and her children's — safety and other needs.

Current "Risk Assessment" tools are often not dynamic enough to deal with all these moving parts.

Further Reading
Bowstead, Janet C. 2021. "Stay Put; Remain Local; Go Elsewhere: Three Strategies of Women's Domestic Violence Help Seeking." *Dignity: A Journal of Analysis of Exploitation and Violence* 6 (3): 4. doi:10.23860/dignity.2021.06.03.04.

https://www.womensjourneyscapes.net/safety-isnt-static/

© Sarah

"Can you blend parts of your life the same way this picture is blended? After all it looks good. But if you look closely you can see that the picture is full of imperfections, just like me and my past"

Sarah

© Sarah

"Knock , knock who dares? What you leave and where you end up. What's behind the doors?"

Sarah

Journey trajectories – time

How long are women and children on the move due to domestic violence?

Even combining years of administrative data often does not fit together completed journeys from abuse to resettlement. Many records start with women in temporary accommodation – refuges, hostels, sleeping rough or staying with friends or family – and end with a relocation to further precarity.

Many factors affect the timing of seeking help. Women may or may not have an option about when and where to seek help – and if they try and access a service they may not find any space at a refuge, or may be put on a waiting list for an advice and support worker. Individual women may seek help at a time of extreme danger – or at a time of opportunity: it may be when a woman hears about support services or refuges, or is encouraged to believe that someone will help her.

Women and children may be on the move for years – including stays that may be hoped to be permanent, but turn out to be only interim – and resettlement may remain provisional.

© Cordelia

© Daisy

Further Reading
Bowstead, Janet C. 2017. "Women on the Move: Theorising the Geographies of Domestic Violence Journeys in England." *Gender, Place and Culture* 24 (1): 108–121. doi:10.1080/0966369X.2016.1251396.

© Amy

"That's a hummingbird – part of a bigger mural and I'd walked past it many times before but I hadn't noticed the hummingbird"

© Amy

"I thought when I saw it the first time that it's beautiful and strong but small"

Amy

Journey trajectories – multiple displacements

Women and children's journeys to escape abuse are often complex; including strategies of staying put, remaining local and going elsewhere, leading to multi-stage trajectories over time and geography.

From initially staying put, both the behaviour of the abuser and the support (or lack of support) of services and authorities may then force relocation. Women's help-seeking strategies may mean that they get the support and protection they need – involving a range of different services – or they might encounter closed doors, judgement and prejudice, lack of belief, misunderstanding, and service responses that make things worse.

Complex, segmented journeys are therefore often made more fragmented by aspects of policy and practice. Not only could more effective actions to hold abusers to account help prevent the need for relocation and/or multiple moves, but further moves and pressure points are also caused by actions or inactions of agencies, and by administrative boundaries. The ongoing displacement – and consequent dislocation – is considerable.

© Shalom

© Shalom

Further Reading
https://www.womensjourneyscapes.net/the-complexity-of-womens-journeys/

Bowstead, Janet C. 2017. "Segmented Journeys, Fragmented Lives: Women's Forced Migration to Escape Domestic Violence." *Journal of Gender-Based Violence* 1 (1): 43–58. doi:10.1332/239868017X14912933953340.

Bowstead, Janet C. 2021. "Stay Put; Remain Local; Go Elsewhere: Three Strategies of Women's Domestic Violence Help Seeking." *Dignity: A Journal of Analysis of Exploitation and Violence* 6 (3): 4. doi:10.23860/dignity.2021.06.03.04.

© Amy

"I drive – I used to drive a hell of a lot for work just under 14,000 miles a year for the last three years"

Amy

© Amy

"So the options – offering sort of choices for your kind of direction. And all the lights are green!"

Amy

Force and Agency

Women's journeys to escape domestic violence can be understood as an active strategy to achieve safety and as forced migration during which women experience force from the abuser and the impact of agencies and authorities. Force and agency may be conceptualised as poles of a continuum in migration, which recognises the operation of power at a range of scales and in relation to both mobility and immobility. Women and children may be both forced to move and forced to stay.

Spatial theorisations of migration have previously tended not to examine forced and gendered processes in countries like the United Kingdom. Identifying women's domestic violence journeys as a process of gendered forced migration within the United Kingdom does not, therefore, deny degrees of agency in how women use space to achieve safety, or deny how finding some sense of control of place is important in recovering from trauma.

Tens of thousands of women and children relocate due to domestic abuse. Their journeys are hidden and secret, but the administrative data reveal them as criss-crossing the country – see the flow map enclosed – the churn below the surface of an overall lack of net effect.

The initial force from the abuser can be reinforced by service and state responses that do not understand or respond to the complexity and intersectionality of women's experiences and actions as they attempt to navigate from abuse to safety and freedom.

Further Reading

https://www.womensjourneyscapes.net/forced-to-move-—-forced-to-stay/

Bowstead, Janet C. 2015. "Forced Migration in the United Kingdom: Women's Journeys to Escape Domestic Violence." *Transactions of the Institute of British Geographers* 40 (3): 307–320. doi:10.1111/tran.12085.

Bowstead, Janet C. 2017. "Women on the Move: Theorising the Geographies of Domestic Violence Journeys in England." *Gender, Place and Culture* 24 (1): 108–121. doi:10.1080/0966369X.2016.1251396.

© Daisy

"Cut off in its prime..."
Daisy

© Marita

"I had no value whatsoever!"
Marita

© Marita

© Marita

Pressure Points

Women often emphasise that the disruption of their journeys does not simply relate to distance, or number of moves, but also to practical and emotional pressure points regardless of mileage or time. These pressure points make journeys more fragmented and disruptive, but they are also aspects that could be addressed more effectively by policy and services. They are opportunities for more effective responses.

Examples of pressure points are a lack of information on rights and services, and a lack of continuity of service provision (for example eligibility restrictions of time or location) and the resulting cliff edges of support and loss of momentum for the journey. Other pressures are the loss of possessions – causing both cost and upset – and the loss of rights to access services and particularly housing across administrative boundaries. What may be, for the woman, the next small step in an ongoing journey is thereby made into a severe cut-off, and therefore far more fragmented and difficult.

In terms of information, the complexity of women's domestic violence journeys means that a wide range of frontline agencies could outline 'route maps' to women, rather than just focusing on delivering their specific responsibilities. An understanding of the long-term nature of women's journeys also highlights that women need more than just a map. Even if women have a sense of where they could go initially, they also need practical and emotional support – the equivalent of road signs and roadside assistance – to help them continue to navigate through unfamiliar areas, and provide continuity through difficult junctions.

© Shalom

Further Reading
Bowstead, Janet C. 2017. "Segmented Journeys, Fragmented Lives: Women's Forced Migration to Escape Domestic Violence." *Journal of Gender-Based Violence* 1 (1): 43–58. doi:10.1332/239868017X14912933953340.

"Different pieces can complete a puzzle – Everyone has a space somewhere"

Sarah

"Supported"

Marita

© Cordelia

"I love my flat – but I hate it because it's like sometimes – especially in summer – it feels like it's a box because I don't have any outdoor spaces or a balcony or anything. So sometimes I feel quite trapped in there"

Cordelia

"Holland Park - Peaceful place to gather all my thoughts and take a break"

Sarah

© Sarah

Processes

State Duties – Admin data as evidence

Women on the move due to domestic violence are invisible to the favoured evidence used to calculate prevalence and make decisions on domestic violence issues and provisions. Anyone on the move or in temporary accommodation is excluded from the sampling frame of extensively used surveys such as the Crime Survey for England and Wales.

So the women most directly and acutely affected by domestic abuse in the past year have no chance of being asked the Crime Survey questions. They have no chance of being counted in this measure of domestic abuse prevalence.

As a result, you might expect that the state would be concerned to remedy this invisibility – to ensure and invest in the generation, availability and use of data to begin to fill this serious gap. However, administrative data from services – both statutory and voluntary – may record domestic violence in an incompatibly wide range of ways; and is not consistently collected, de-identified or archived to be used as evidence. It is not that the evidence doesn't exist – there is extensive monitoring of contracted services – but that it is being neglected, withheld and destroyed.

You might almost think that Government does not want to know the extent of what is going on in terms of domestic violence help-seeking – and the extent of the mismatch between the scale of provision and the scale of need.

Further Reading

Bowstead, Janet C. 2019. "Women on the Move: Administrative Data as a Safe Way to Research Hidden Domestic Violence Journeys." *Journal of Gender-Based Violence* 3 (2): 233–248. doi:10.1332/239868019X15538575149704.

Bowstead, Janet C., Stuart Hodkinson, and Andy Turner. 2020. "Uncovering Internally Displaced People in the Global North through Administrative Data: Case Studies of Residential Displacement in the UK." In *The Handbook of Displacement*, edited by Peter Adey, Janet C. Bowstead, Katherine Brickell, Vandana Desai, Mike Dolton, Alasdair Pinkerton, and Ayesha Siddiqi, 431–450. London: Palgrave Macmillan. https://doi.org/10.1007/978-3-030-47178-1_30.

Bowstead, Janet C. (forthcoming). "Journeyscapes: the regional scale of women's domestic violence journeys."

"I felt like I'd just been rescued. So like my lifebuoy would be Solace, and family, and people who helped to rescue me from it all. Because of the support system I had, even though they were far, I was able to leave. So I'm very grateful. And I didn't go back. Because I think, if I didn't have the support system I had, then…I don't know. Because, it's easy to go back, you know. Even though it's horrible, they have so wired you up to a point that you don't even know what – you can't even distinguish between bad treatment or not. I've done a lot of reading and it's opened my mind to understand how these people work – controlling of everything"

© Sha

© Shalom

"When I saw this anchor I thought – wow – because for me it symbolises a lot of things – an anchor – because you need to be anchored in the positive things and hold tight. And support is also like an anchor; so it's very metaphorical"

© Shalom

"And this is Danger – this is like a Warning – Don't go back"

Shalom

State Duties – Journeyscapes

All too often, beyond the original escape, women's domestic violence journeys continue to be fragmented and disorientating over both time and space. Women have little control over their mobility – where they go, how long they stay in temporary accommodation, whether they have to keep on moving.

This can be contrasted with the concept of a functional scale for domestic violence journeys – "journeyscapes" – whereby women and children travel as far as they need to escape the abuse, but are not forced any further due to administrative boundaries or services. A society which thinks and responds more coherently in terms of policy, services and rights could journeyscape women's experiences and help them re-establish control over their sense and reality of home.

The responsibilities on the state would be to journeyscape:
> Through effective policies, laws, professional practice, and awareness
> To build the infrastructure and map the terrain
> To minimise the losses so women and children retain their rights and status
> But not to determine the route that any woman takes

The principle should be that women – and their children – go as far as they need / stay as near as they can; and have the right to a life free from abuse.

---

journeyscape (ˈdʒɜːniˌskeɪp) *n.*

1. a travelling from one place to another when viewed as a whole from a single aspect.

2. the scene or view of a journey, especially a coherent representation of what might otherwise appear fragmented.

3. a. the overall view of the distance travelled in a journey.
   b. the overall view of the time taken to make a journey.

journeyscape (ˈdʒɜːniˌskeɪp) *vb. (tr.)*

1. to improve the existing features of a journey.

Just as an area of land begins to make more sense – and feel safer and more navigable – when it is seen as a landscape; this research explores how a forced journey away from domestic abuse could become more positive when seen and experienced in the context of a journeyscape.

"This really really long avenue"

Kelly

"This picture is like a door – and you want to go somewhere – and this is your future"

Carol

"To get to the end – yes – as things come together you find the path you can go. And you get to the point where that one person is – standing, in the light"

Shalom

State Duties – Laws

Laws are obviously part of the means by which Governments carry out their duties, allocate duties to other scales of administration (such as Local Government), and specify the behaviours of individuals and authorities. They can enshrine rights, specify norms, and provide routes to civil, family or criminal justice. They could be part of the state stepping up to its duties towards women and children on the move – part of journeyscaping the currently hazardous and hostile terrain faced by women trying to escape abuse.

But laws are nowhere near enough, and may not even be the most important way forward. Governments often find it easier to pass another law which claims to tackle a complex issue, rather than implement the laws that are already in place. And rather than develop and sustain the services and social infrastructure that are needed to make a real difference.

Many women who experience abuse go nowhere near the criminal justice system – often these are the tens of thousands of women (and children) who relocate every year in the UK. If the Government wants to change the options and life chances of these women and children, it needs to ensure everything from lessons on healthy relationships in schools to sustained funding for the specialist services all around the country that enable woman and children to rebuild their lives after abuse.

Rather than focusing primarily on new powers to the criminal justice system, led by the Home Office or the Ministry of Justice, this would need effective action across the board by ministries such as those responsible for housing, communities, local government, education, health and social care, work and pensions. There has never been a law *against* any of that: never been a law stopping effective action and the long-term work of developing a society that prevents abuse, supports survivors and builds real lasting justice.

Further Reading
https://www.womensjourneyscapes.net/do-we-need-another-law/
https://www.womensjourneyscapes.net/who-smelt-it-dealt-it/

FOAM

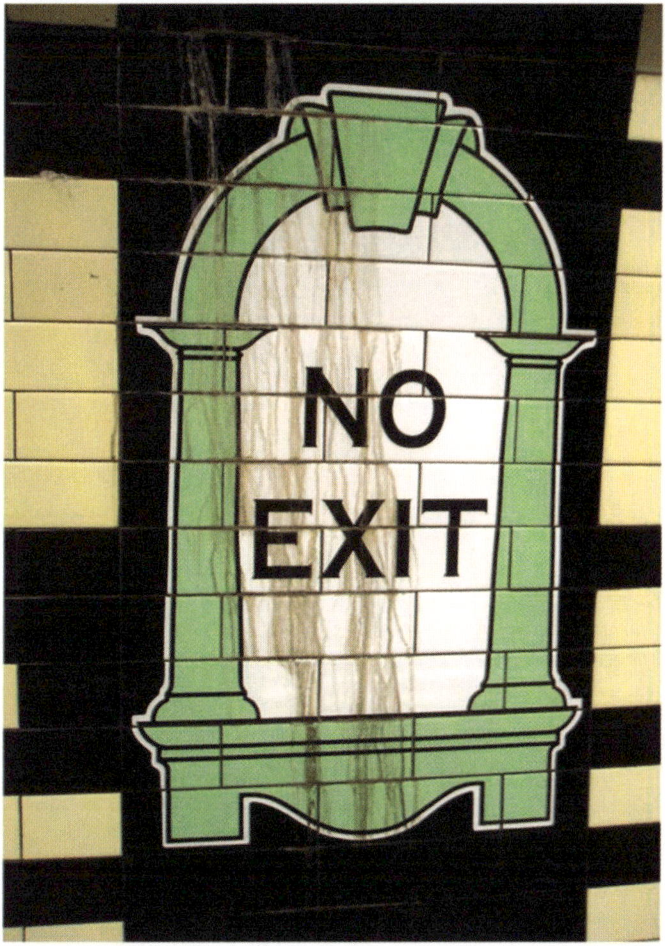
NO
EXIT

PRESS
EMERGENCY
DOOR RELEASE

FIRE
BLANKET

FIR
EXI

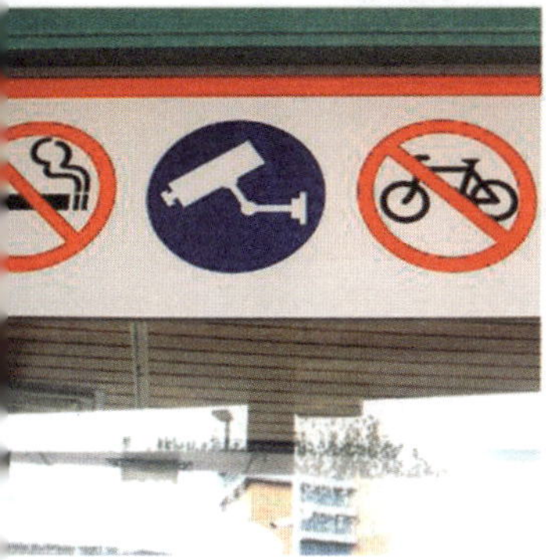

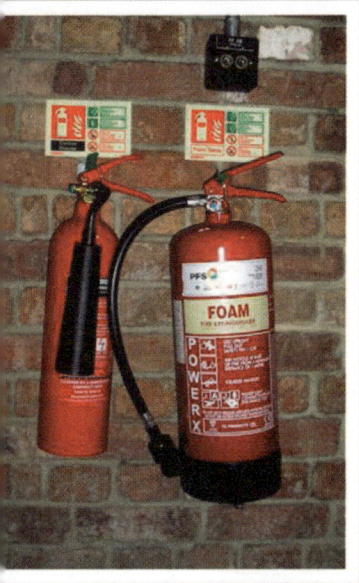

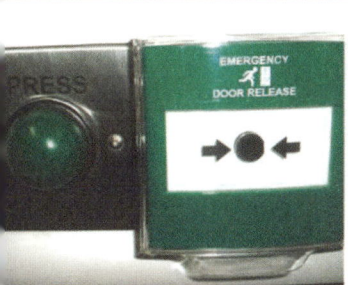

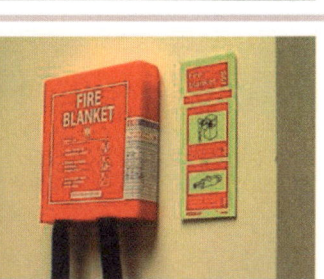

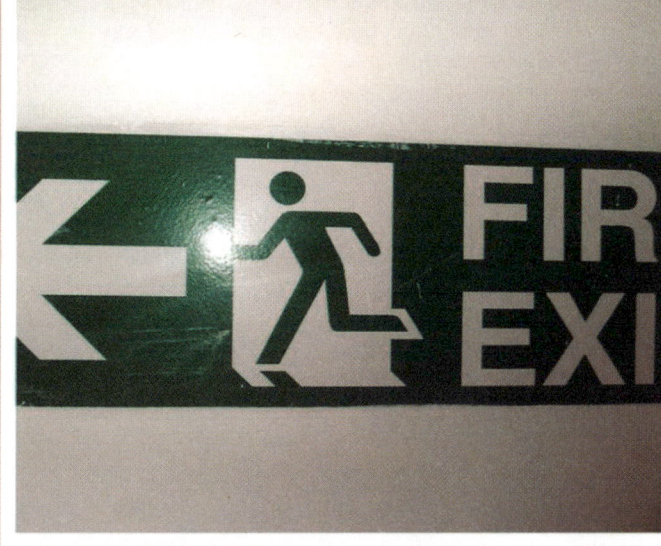

© Amy

"Signs and signals on the Underground. I was thinking about how much I hid for so long and didn't signal anything and then once I did – it wasn't enough"

Amy

State Duties – IDPs

According to the data it provides to the United Nations, the UK has no Internally Displaced Persons (IDPs) or people in IDP-like situations (persons displaced by armed conflict, generalised violence and human rights violations). Journeys of internal migration in the UK tend to be seen as options that people take for financial, housing, education or employment reasons.

However, women's forced internal displacement due to domestic abuse *is* due to a human rights violation – creating clear duties on the state to address their particular needs: *"Unlike refugees, the internally displaced have not left the country whose citizens they normally are. As such, they remain entitled to the same rights that all other persons in their country enjoy. They do, however, have special needs by virtue of their displacement."*[1]

Imagine a world where you could escape if you needed to – and therefore also knew that you could try and stay put (using legal protection and services' support if necessary) if you wanted to…. because there would always be a safety net if that didn't work.

Imagine a support system that gave top priority to your rights and needs – that was there to serve you: a basic infrastructure that you have a fundamental right to access – when and where you need it.

Not the current fragmentation – where local areas can decide whether or not to provide services – and restrict the services they provide to local women and children. Where you have to keep on proving that you really need help, and prove where you have come from – as if you are asking for some special favour rather than simply your rights…

If we really believe that *"violence against women constitutes a violation of the rights and fundamental freedoms of women and impairs or nullifies their enjoyment of those rights and freedoms"*[2] then every state should ensure that women can easily access their right to escape violence. Every state should provide a comprehensive, fully-functioning infrastructure for women's human rights; including the state's duties towards IDPs – women and children on the move due to domestic violence.

References
[1] *Handbook for applying the Guiding Principles on Internal Displacement* (UN Office for the Coordination of Humanitarian Affairs, 1999, p. 5) http://www.unhcr.org/en-us/protection/idps/50f94df59/handbook-applying-guiding-principles-internal-displacement-ocha-november.html
[2] UN General Assembly. 1993. Declaration on the Elimination of Violence against Women: General Assembly resolution 48/104 of 20 December 1993. United Nations, Geneva, Switzerland. http://www.un-documents.net/a48r104.htm

"I've got all my bits and pieces on there – it's nice, and it's all lit up at night time – My little cosy corner"

Kate

"And this is my bedside – I've got some flowers – from donations yesterday!"

Shalom

"My Bedroom. The photo of my parents who passed away, my Rosary and Little Jesus and the Light"

Lulu

© Daisy

"I like the wires there –
the lines of communication"

Daisy

© Daisy

© Marilyn

"The hidden mess
– how the mess is
normally hidden"

Marilyn

"It's about looking
up... because that
was on the top of
a high wall, and
normally I'm
always looking
down"

© Marilyn

Marilyn

Strategies - Scale

Despite such disruption at the individual scale, domestic violence journeys do not aggregate into net migration flows at the local or national scale and the overall process is one of spatial churn. The multi-scale analysis of this research reveals the turbulence beneath the surface.

Scale matters. Just because needs are expressed locally does not mean that they are *caused* locally, by isolated individual-scale issues – or that the responses should be devised and provided in local silos.

Domestic abuse affects the whole country – the whole society – so it is vital that those in power make decisions and provide responses at the right scale: recognising what must be national – what local – and understand the serious consequences for getting this right or wrong.

A lack of understanding of the scale of women's and children's domestic violence journeys – and the service needs that arise from them – has led to an incorrect analysis of the scale required for effective service responses. And this continues: Part 4 of the Domestic Abuse Act 2021 devolves responsibility for both assessing and providing for the need for safe accommodation to Tier 1 local authorities (Counties and Unitary – and London). It's a risky strategy of government: If Local Authorities underestimate cross-border needs for domestic abuse services, there isn't going to be anywhere for women and children to go.

Further Reading

Bowstead, Janet C. 2015. "Forced Migration in the United Kingdom: Women's Journeys to Escape Domestic Violence." *Transactions of the Institute of British Geographers* 40 (3): 307–320. doi:10.1111/tran.12085.

Bowstead, Janet C. 2015. "Why Women's Domestic Violence Refuges Are Not Local Services." *Critical Social Policy* 35 (3): 327–349. doi:10.1177/0261018315588894.

Bowstead, Janet C. (forthcoming). "Journeyscapes: the regional scale of women's domestic violence journeys."

https://www.womensjourneyscapes.net/the-scale-of-services/

© Sarah

"So after all, I had my boys and nothing else. I felt blessed in so many ways because I had my boys with me, but when they fell asleep I was forever counting the patterned circles .. 89, 90, 100 ... "

Sarah

© Cordelia

"Down at the Southbank – it said love is the only language I speak fluently – and then underneath it says void and it's kind of like yeah, I get that"

Cordelia

Strategies – Boundaries

In housing law in England it has long been recognised that to escape domestic violence you may need to leave home, and travel quite a distance – including across local authority boundaries. Access to social housing would usually require a 'local connection' to that local authority – like a moat created between each council area. However, an exception exists for individuals who are unable to remain safely in their own local authority – like a drawbridge extended across these moats – creating a route to safety. So women escaping domestic violence are able to apply to a local authority where they have no 'local connection'.

But these drawbridges are being pulled up in all kinds of ways – cutting off escape routes for women and children. As these drawbridges are closed to women who need to relocate to another area, their escape journeys are made more fragmented – more risky, costly and disruptive – by law, policy and practice.

In addition, data are often only collected within administrative boundaries. So authorities which devise policies, and plan and commission services, only have evidence of what happens within their boundaries. They do not know the impact and needs beyond their borders – including their women and children forced to leave. Administrative lines on the map become cliff edges of knowledge – and cliff edges of provision and support.

It doesn't have to be organised like this. Women and children who relocate across boundaries due to domestic abuse should not see their lives disappearing through the gaps in policy, service provision and rights.

> **Further Reading**
> Bowstead, Janet C. 2021. "Stay Put; Remain Local; Go Elsewhere: Three Strategies of Women's Domestic Violence Help Seeking." *Dignity: A Journal of Analysis of Exploitation and Violence* 6 (3): 4. doi:10.23860/dignity.2021.06.03.04.
>
> Bowstead, Janet C. (forthcoming). "Journeyscapes: the regional scale of women's domestic violence journeys."
>
> https://www.womensjourneyscapes.net/pulling-up-the-drawbridge/
>
> https://www.womensjourneyscapes.net/data-boundaries-knowledge-boundaries/

© Sarah

"A small Picture with a big meaning. A smile for all – be the reason someone smiles today"

Sarah

© Daisy

© Daisy

Strategies - Referrals and Support Services

Each woman has been dealing with the reality and consequences of domestic abuse from before she has any contact with services; and will be doing so for long afterwards. But services have key roles in both referring women and children to support; and in providing that support.

Women are experts in their own lives, and are passing through a complicated and fragmented system which may or may not help them. At each encounter with services they may need simply to be *allowed* to continue their journey, they may need to be *enabled* in practical or emotional ways, or they may need significant and specialist *assistance*.

Women refer themselves to services in about a quarter of cases where the referrer is recorded and are referred by voluntary sector services in another quarter of cases; with statutory services being involved in referring half of the women accessing services. Different statutory agencies (such as Police, Social Services, Housing and Health) are involved to different extents, and there are significantly different patterns of referrals in terms of both women's strategies, and the types of services accessed.

As a result, different types of services (statutory and voluntary) have quite different roles – and different understandings – not least because they interact with women using quite different strategies.

© Marilyn

Further Reading

Bowstead, Janet C. 2021. "Stay Put; Remain Local; Go Elsewhere: Three Strategies of Women's Domestic Violence Help Seeking." *Dignity: A Journal of Analysis of Exploitation and Violence* 6 (3): 4. doi:10.23860/dignity.2021.06.03.04.

https://www.womensjourneyscapes.net/the-opportunity-to-allow-enable-or-assist/

© Amy

"The sun shines here every day – even when you can't see it yet

The sun is shining – above the clouds"

Amy

"The welcoming pineapple – these little mementos of colonial history – you will see them all over at the ends of stairways on fences and all of these things.
It was a brutal time period but the idea of a place being welcoming or wanting to welcome you is what I was thinking about"

Amy

© Amy

Strategies – Women's Refuges

Women's refuges in England were developed as part of a wider feminist movement with goals in terms of freedom, autonomy and equality; of which responses to end male violence were an aspect. Autonomous women's refuges therefore aim to tackle the causes as well as the consequences of domestic abuse, and create safe spaces for women's empowerment.

Since the 1970s there have been women's domestic violence refuges providing temporary accommodation, and a range of support services, in many local authorities. Women's refuges are distinctive services that enable women and their children to relocate to escape domestic violence; and enable them to relocate to a place where they do not have an existing local connection. Accessing a refuge does not mark the end of their journeys away from violence, but is an important stage in that process, and it is therefore vital that they are available in all types of places, across the country.

Refuges in England are currently under threat, not just from years of cuts to publicly-funded services, but from the very specific mis-match between the scale of funding and the scale of need. Part 4 of the Domestic Abuse Act 2021 combines women's refuges with other types of accommodation services – without a proper recognition of the very different role of refuges compared to other accommodation services in women and children's help-seeking. And, crucially, it devolves the responsibility for provision to local authorities – which is not the predominant scale at which women and children need to access refuge services.

Imagine instead a world where there were enough refuge spaces for women and children escaping domestic violence. Where these refuges were situated all around the country – in all types of places – so that women could go to the right type of place for their needs.
- Not too close – first and foremost you need safety
- Not too familiar – the abuser might try and track you down
- Not too far – so that you don't feel that you have been forced into exile
- Not too strange – the kind of place where you can start again

Further Reading

Bowstead, Janet C. 2015. "Why Women's Domestic Violence Refuges Are Not Local Services." *Critical Social Policy* 35 (3): 327–349. doi:10.1177/0261018315588894.

Bowstead, Janet C. 2019. "Safe Spaces of Refuge, Shelter and Contact: Introduction." *Gender, Place and Culture* 26 (1): 52–58. doi:10.1080/0966369X.2019.1573808.

© Shalom

"Cooking together in the refuge: We didn't plan it – it was a very random thing; but I think we're going to plan it next time for the house. I went to show them Brixton and then we went into the shops, and they wanted fruit, and then we ended up seeing the crabs – crabs! – so we got the crabs, and then the other things – and cooked it together"

Shalom

Further Reading

Bowstead, Janet C. 2019. "Spaces of Safety and More-than-Safety in Women's Refuges in England." *Gender, Place and Culture* 26 (1): 75–90. doi:10.1080/0966369X.2018.1541871.

https://www.womensjourneyscapes.net/wp-content/uploads/2018/11/Womens-Journeyscapes-Briefing-paper-4-November-2018.pdf

https://www.womensjourneyscapes.net/wp-content/uploads/2019/06/Womens-Journeyscapes-Briefing-paper-5-June-2019.pdf

https://www.womensjourneyscapes.net/cross-border-support/

© Violet

"Images from the refuge where I like to relax and how was I settling in"

Violet

© Violet

© Kelly

"Positive vibes, Positive attitude – I know what makes me smile. What makes you smile?"

© Sarah

Sarah

"So these are my things – some are cheap, some are more money – to treat myself"
© Carol

Carol

© Kate

© Kate

"That was my bath time – just lay in the bath and chill. My little speaker – it has a lot of bass on it, and I put it on the bath, and the whole water was moving – and I thought I better turn it down!"

"I enjoy cooking – that's like tuna, onion, garlic, tomato, chilli flakes, noodles and broccoli. I like to cook all the spicy food that the kids won't eat – they just want like chicken nuggets"

Kate

© Favour

© Lulu

© Lee

Strategies – Housing and Home

'Home' is always a powerful concept – the 'domestic' in domestic abuse/domestic violence highlights the complexity of what home means. There is a risk that the focus on women and children escaping to a safe place implies that the escape phase is the only critical stage in domestic violence journeys.

Women and children on the move are making and remaking home – as much as they can – in all kinds of temporary accommodation; as well as hoping to re-establish a home on a more secure basis. But many women find that their housing security is reduced by what they had to do for their personal security; with owner-occupiers ending up as private renters, and women with secure social housing tenancies ending up in short-term accommodation with no hope that any housing will be 'permanent' again.

When you need to relocate because of domestic abuse there can be so many losses – from your sense of home to your personal possessions, from your career progression to your comfy sofa, from your favourite corner shop to your children's friends, from your degree certificate to your cats. It often depends on how you have to leave – whether the abuser will notice any plans you make, anything you try to do to prepare – and on what kind of help and support you get along the way: practical, legal, financial, emotional.

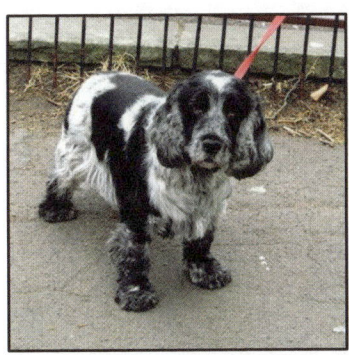
© Marilyn

© Daisy

Further Reading
Bowstead, Janet C. 2021. "Stay Put; Remain Local; Go Elsewhere: Three Strategies of Women's Domestic Violence Help Seeking." *Dignity: A Journal of Analysis of Exploitation and Violence* 6 (3): 4. doi:10.23860/dignity.2021.06.03.04.

https://www.womensjourneyscapes.net/womens-struggle-to-be-free/

Formula for service provision

Administrative data, as used in this research, record formal help-seeking: only the women and children who actually accessed support services, and not those who didn't seek or receive such help. This therefore represents expressed need, rather than hidden need.

From this research, a formula for different types of services provision in England for women – recognising how women use refuges, other types of accommodation services, and non-accommodation services – has calculated the minimum provision required for this help-seeking due to domestic abuse. It uses administrative records of women's help-seeking to services; as well as data on service location and capacity, and on characteristics of people and places, to analyse their association (or not) with different strategies and rates.

Further Reading
https://www.womensjourneyscapes.net/formula-for-service-provision/

Formula for service provision

Accommodation services:
A minimum of 5,369 family bedspaces, of which 4,497 should be 'Women's Refuge' spaces and 872 'Other' types of support accommodation. Women's refuge provision must include specialist 'by and for' provision, in addition to being women-only; whereas other accommodation may be more generic but equally may be for higher and specialist support needs, such as by providing 24 hours' staffing, separate rooms for teenage children, or particular staff specialisms. Staffing roles and levels must therefore be factored in beyond the bricks-and-mortar of 'bedspaces' to provide genuine support capacity.

N.B. The majority of help-seeking to accommodation is across Tier 1 administrative boundaries – 60% – but this is made up of women's different strategies to refuges (65% across boundaries) in comparison to other types of support accommodation (33% across boundaries). Planning, funding and provision – as well as eligibility – must therefore be across these boundaries, at the national and regional scales.

Non-accommodation services:
A minimum of 1,084 fte (full-time-equivalent) community-based specialist support workers (separate roles from 'advice'; or risk-based 'advocacy'); rising to a minimum of 1,543 fte workers to be able to support women with additional needs beyond the domestic abuse. Specialist workers such as outreach, support or resettlement workers will work with a maximum number of women at any one time ('caseload') and for a range of timescales. A rights-based approach would provide holistic support, without arbitrary time limits. From this research, the timescales are based on the actual length of time women received such services, so are very much a minimum. This also does not include at all the support services children need and deserve.

N.B. The vast majority of help-seeking to non-accommodation services is within Tier 1 administrative boundaries, or within London as a region, but access must still be needs- and rights-based and therefore available across boundaries.

Further Reading
https://www.womensjourneyscapes.net/formula-for-service-provision/

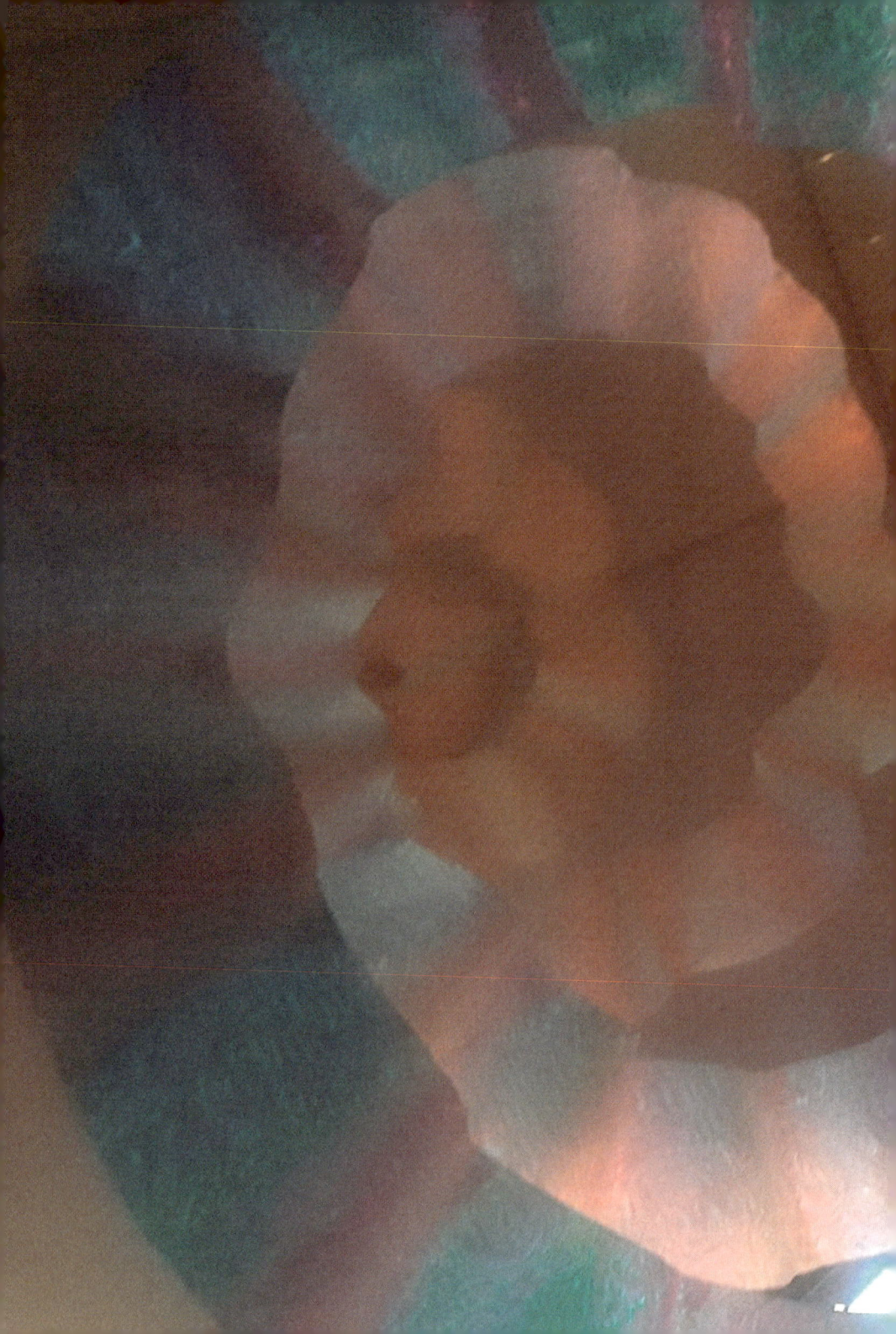